Everyday Mathematics®

The University of Chicago School Mathematics Project

Student Math Journal
Volume 2

Grade

P9-DOC-446

McGraw Hill Education

Chicago, IL • Columbus, OH • New York, NY

The University of Chicago School Mathematics Project (UCSMP)

Max Bell, Director, UCSMP Elementary Materials Component; Director, *Everyday Mathematics* First Edition; James McBride, Director, *Everyday Mathematics* Second Edition; Andy Isaacs, Director, *Everyday Mathematics* Third Edition; Amy Dillard, Associate Director, *Everyday Mathematics* Third Edition; Rachel Malpass McCall, Associate Director, *Everyday Mathematics* Common Core State Standards Edition

Authors
Max Bell, Jean Bell, John Bretzlauf, Amy Dillard, Robert Hartfield, Andy Isaacs, James McBride, Rachel Malpass McCall, Kathleen Pitvorec, Peter Saecker

Technical Art
Diana Barrie

Third Edition Teachers in Residence
Jeanine O'Nan Brownell, Andrea Cocke, Brooke A. North

UCSMP Editorial
Rossita Fernando, Lila K. Schwartz

Contributors
Allison Greer, Meg Schleppenbach, Cynthia Annorh, Robert Balfanz, Judith Busse, Mary Ellen Dairyko, Lynn Evans, James Flanders, Dorothy Freedman, Nancy Guile Goodsell, Pam Guastafeste, Nancy Hanvey, Murray Hozinsky, Deborah Arron Leslie, Sue Lindsley, Mariana Mardrus, Carol Montag, Elizabeth Moore, Kate Morrison, William D. Pattison, Joan Pederson, Brenda Penix, June Ploen, Herb Price, Dannette Riehle, Ellen Ryan, Marie Schilling, Susan Sherrill, Patricia Smith, Kimberli Sorg, Robert Strang, Jaronda Strong, Kevin Sweeney, Sally Vongsathorn, Esther Weiss, Francine Williams, Michael Wilson, Izaak Wirzup

Photo Credits
Cover (l)C Squared Studios/Getty Images, (r)Tom & Dee Ann McCarthy/CORBIS, (bkgd)Ralph A. Clevenger/CORBIS; **Back Cover** C Squared Studios/Getty Images; **others** The McGraw-Hill Companies.

---

everyday**math**.com

STEM McGraw-Hill is committed to providing instructional materials in Science, Technology, Engineering, and Mathematics (STEM) that give all students a solid foundation, one that prepares them for college and careers in the 21st century.

Send all inquiries to:
McGraw-Hill Education
STEM Learning Solutions Center
P.O. Box 812960
Chicago, IL 60681

ISBN: 978-0-07-657639-5
MHID: 0-07-657639-6

Printed in the United States of America.

9 QMD 17 16 15 14 13

Contents

UNIT 6 Developing Fact Power

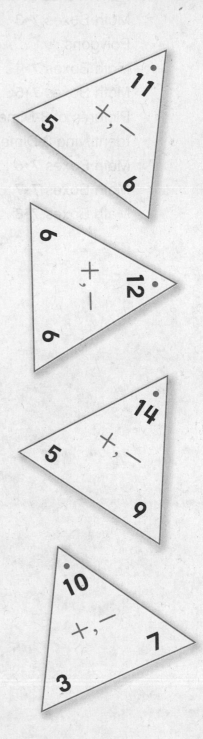

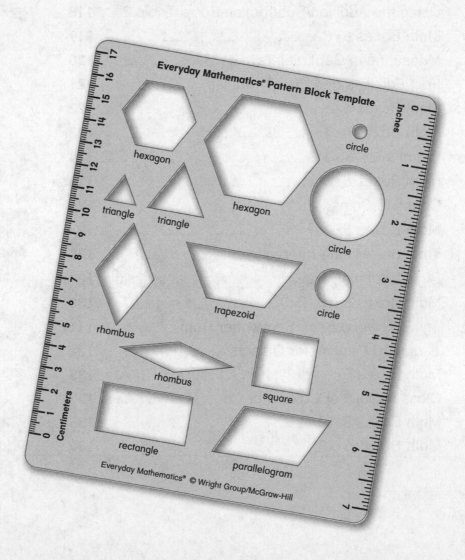

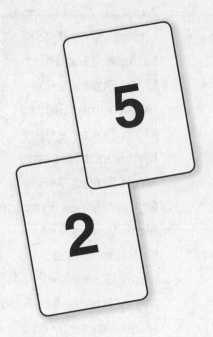

Activity Sheets

LESSON 6·1 Dice-Throw Record 2

Record each fact and its turn-around fact once.

Unit
dice
dots

2	3	4	5	6	7	8	9	10	11	12
			1+4				6+3			
			4+1							

LESSON 6·1 Math Boxes

---

1. Write the sums.

7 + 10 = ____

____ = 4 + 8

____ = 9 + 5

5 + 6 = ____

---

2. What is the missing rule?

in
↓
 Rule

in	out
3	5
17	19
14	16

out

Fill in the circle next to the best answer.

Ⓐ +3   Ⓑ +2

Ⓒ −2   Ⓓ −5

---

3. Add.

1 + 9 = ____

8 + 6 = ____

9 + 9 = ____

10 + 4 = ____

---

4. Draw lines to match the shapes that look alike.

Column A   Column B

LESSON 6·2 — Name-Collection Boxes

1. Write other names for 11.

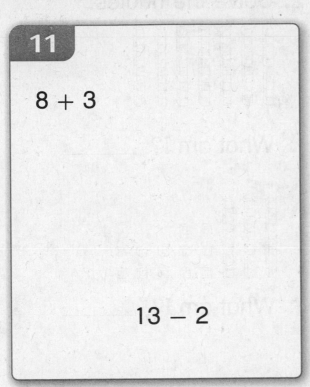

11

8 + 3

13 − 2

2. Write other names for 12.

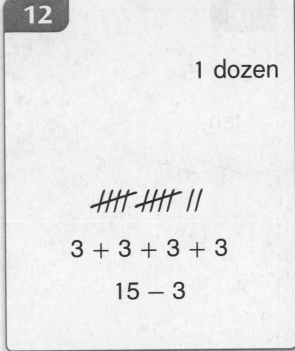

12

1 dozen

‖‖‖ ‖‖‖ ‖‖

3 + 3 + 3 + 3

15 − 3

3. Cross out the names that don't belong in the 10-box.

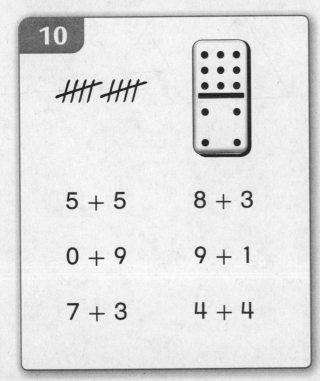

10

‖‖‖ ‖‖‖

5 + 5   8 + 3

0 + 9   9 + 1

7 + 3   4 + 4

4. Make your own.

Math Boxes

1. Write 5 more names for 10.

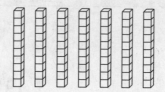

5 + 5

ten

~~HHt~~ ~~HHt~~

2. Solve the riddles.

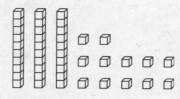

What am I? _____

What am I? _____

3. Add.

2 + 2 = _____

3 + 3 = _____

4 + 4 = _____

5 + 5 = _____

4. Shade the biggest triangle.

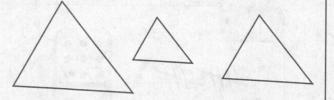

Date _____

Write the 3 numbers for each domino.
Use the numbers to write the fact family.

1.

Numbers: _____, _____, _____

Fact Family: ____ + ____ = ____

____ + ____ = ____

____ − ____ = ____

____ − ____ = ____

2.

Numbers: _____, _____, _____

Fact Family: ____ + ____ = ____

____ + ____ = ____

____ − ____ = ____

____ − ____ = ____

3.

Numbers: _____, _____, _____

Fact Family: ____ + ____ = ____

____ + ____ = ____

____ − ____ = ____

____ − ____ = ____

4. Make up your own.
Draw the dots.

Numbers: _____, _____, _____

Fact Family: ____ + ____ = ____

____ + ____ = ____

____ − ____ = ____

____ − ____ = ____

LESSON 6·3 Name-Collection Boxes

Write as many names as you can for each number.

1.

13

2.

20

3. Cross out the names that don't belong in the 25 box.

4. Choose a number. Show as many names for it as you can.

25

$5 + 5 + 5 + 5$

$25 + 0$ $17 + 18$

$30 - 5$ $24 + 1$

$19 + 4$

Date _____

1. Write the sums.

6 + 10 = _____

_____ = 8 + 7

_____ = 3 + 9

6 + 7 = _____

2. Fill in the missing numbers.

in ↓

Rule

+10

out ↓

in	out
13	
18	
79	
93	
125	

3. Add.

9 + 8 = _____

5 + 7 = _____

6   10
+ 4   + 9

4. Draw lines to match the shapes that look alike.

Column A   Column B

LESSON 6·4 Fact Power Table

0 +0 *0*	0 +1 *1*	0 +2 *2*	0 +3 *3*	0 +4 *4*	0 +5 *5*	0 +6 *6*	0 +7 *7*	0 +8 *8*	0 +9 *9*
1 +0 *1*	1 +1 *2*	1 +2 *3*	1 +3 *4*	1 +4 *5*	1 +5 *6*	1 +6 *7*	1 +7 *8*	1 +8 *9*	1 +9 *10*
2 +0 *2*	2 +1 *3*	2 +2 *4*	2 +3 *5*	2 +4 *6*	2 +5 *7*	2 +6 *8*	2 +7 *9*	2 +8 *10*	2 +9 *11*
3 +0	3 +1	3 +2	3 +3	3 +4	3 +5	3 +6	3 +7	3 +8	3 +9
4 +0	4 +1	4 +2	4 +3	4 +4	4 +5	4 +6	4 +7	4 +8	4 +9
5 +0	5 +1	5 +2	5 +3	5 +4	5 +5	5 +6	5 +7	5 +8	5 +9
6 +0	6 +1	6 +2	6 +3	6 +4	6 +5	6 +6	6 +7	6 +8	6 +9
7 +0	7 +1	7 +2	7 +3	7 +4	7 +5	7 +6	7 +7	7 +8	7 +9
8 +0	8 +1	8 +2	8 +3	8 +4	8 +5	8 +6	8 +7	8 +8	8 +9
9 +0	9 +1	9 +2	9 +3	9 +4	9 +5	9 +6	9 +7	9 +8	9 +9

LESSON 6·4 **Math Boxes**

1. Label the box.
Add 5 names.

20 − 5

7 + 8

2. What is the number?

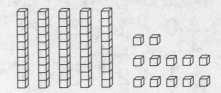

Fill in the circle next to the best answer.

Ⓐ 37

Ⓑ 62

Ⓒ 512

Ⓓ 58

3. Add.

3 + 0 = _____

3 + 1 = _____

$$\begin{array}{r} 6 \\ +\ 0 \\ \hline \end{array} \qquad \begin{array}{r} 6 \\ +\ 1 \\ \hline \end{array}$$

4. Shade the biggest circle.

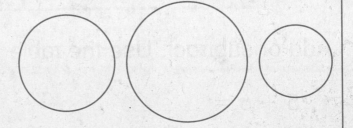

Date _____

LESSON 6·5

Using the Addition/Subtraction Facts Table 2

+,−	0	1	2	3	4	5	6	7	8	9
0	0	1	2	3	4	5	6	7	8	9
1	1	2	3	4	5	6	7	8	9	10
2	2	3	4	5	6	7	8	9	10	11
3	3	4	5	6	7	8	9	10	11	12
4	4	5	6	7	8	9	10	11	12	13
5	5	6	7	8	9	10	11	12	13	14
6	6	7	8	9	10	11	12	13	14	15
7	7	8	9	10	11	12	13	14	15	16
8	8	9	10	11	12	13	14	15	16	17
9	9	10	11	12	13	14	15	16	17	18

Add or subtract. Use the table to help you.

1. $5 + 6 =$ _____

2. $11 - 5 =$ _____

3. $8 + 4 =$ _____

4. $12 - 4 =$ _____

5. $7 + 8 =$ _____

6. $15 - 8 =$ _____

7. $9 + 9 =$ _____

8. $18 - 9 =$ _____

9. $9 + 7 =$ _____

10. $16 - 9 =$ _____

Date _____

1. Subtract.

5 − 1 = _____

4 − 2 = _____

_____ = 6 − 0

_____ = 3 − 3

2. Write the fact family.

_____ + _____ = _____

_____ + _____ = _____

_____ − _____ = _____

_____ − _____ = _____

3. Use your number grid.

Start at 31.

Count up 19.

You end at _____.

31 + 19 = _____

4. Draw lines to match the shapes that look alike.

Column A   Column B

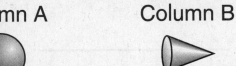

LESSON 6·6 — Measuring in Centimeters

1. Use 2 longs to measure objects in centimeters. Record their measures in the table.

Object (name it or draw it)	My measurement
	about _____ cm
	about _____ cm

Use a ruler to measure to the nearest centimeter.

2. _____

   about _____ cm

3. _____

   about _____ cm

4. _____

   about _____ cm

5. _____

   about _____ cm

6. Draw a line segment that is about 9 centimeters long.

Date _____

Write the fact family for each Fact Triangle.

1.

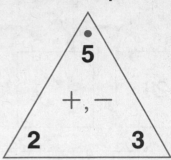

___ + ___ = ___

___ + ___ = ___

___ − ___ = ___

___ − ___ = ___

2.

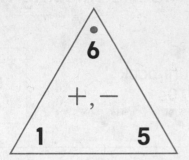

___ + ___ = ___

___ + ___ = ___

___ − ___ = ___

___ − ___ = ___

3.

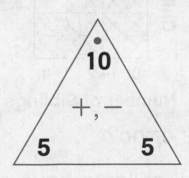

___ + ___ = ___

___ + ___ = ___

___ − ___ = ___

___ − ___ = ___

4. Write the missing number.
   Write the fact family.

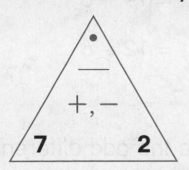

___ + ___ = ___

___ + ___ = ___

___ − ___ = ___

___ − ___ = ___

LESSON 6·6 **Math Boxes**

1. Write the missing numbers.

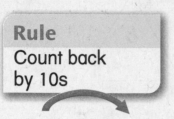

Rule
Count back
by 10s

| 44 | 34 | 24 | | |

2. Make a pattern. Use your Pattern-Block Template.

3. Subtract.

$11 - 8 =$ _____

_____ $= 14 - 8$

_____ $= 12 - 6$

$10 - 7 =$ _____

Circle the odd differences.

4.

Number of Siblings

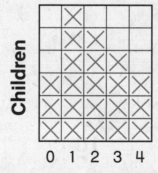

Children

0 1 2 3 4

Number of Siblings

Yes or no?

Do 4 children have exactly

2 siblings? _____

Do the greatest number

of children have just

1 sibling? _____

Date _____

1. Subtract.

$4 - 1 =$ _____

$7 - 2 =$ _____

$$\begin{array}{r} 5 \\ -\ 4 \\ \hline \end{array} \qquad \begin{array}{r} 6 \\ -\ 0 \\ \hline \end{array}$$

2. Write the fact family.

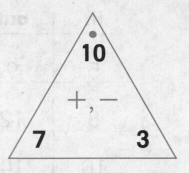

_____ + _____ = _____

_____ + _____ = _____

_____ − _____ = _____

_____ − _____ = _____

3. Use your number grid.

Start at 36. Count back 14.

$36 - 14 =$ _____?

Fill in the circle next to the best answer.

(A) 50

(B) 29

(C) 18

(D) 22

4. Draw lines to match the shapes that look alike.

Column A   Column B

LESSON 6·8   "What's My Rule?"

1. Find the rule.

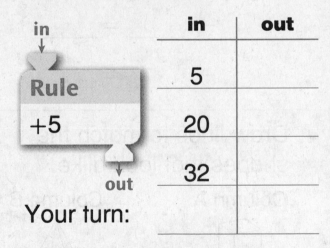

in

Rule ▢

out

in	out
5	9
8	12
10	14

Your turn:

2. Fill in the blanks.

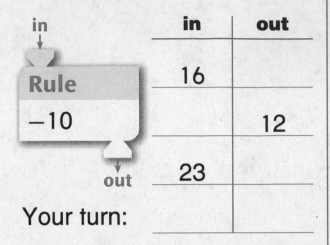

in

Rule
−10

out

in	out
16	
	12
23	

Your turn:

3. What comes out?

in

Rule
+5

out

in	out
5	
20	
32	

Your turn:

4. What goes in?

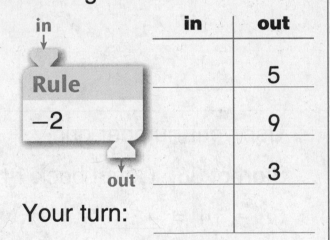

in

Rule
−2

out

in	out
	5
	9
	3

Your turn:

Make up your own.

5.

in

Rule ▢

out

in	out

6.

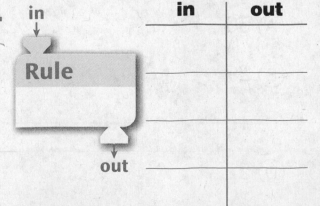

in

Rule ▢

out

in	out

Date _____

Math Boxes

---

1. Write the missing numbers.

 **Rule**
−2

32		28	26	

---

2. Draw the next two figures.

      _____ _____

---

3. Subtract.

18 − 10 = _____

_____ = 14 − 9

_____ = 16 − 8

10 − 3 = _____

Circle the even differences.

4. Spinner Results

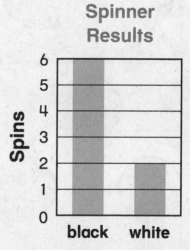

How many more times did the spinner land on black than on white?

Fill in the circle next to the best answer.

Ⓐ 6   Ⓑ 2   Ⓒ 8   Ⓓ 4

LESSON 6·9 Counting Coins

 1¢
$0.01
a penny

 5¢
$0.05
a nickel

 10¢
$0.10
a dime

Q 25¢
$0.25
a quarter

How much money? Use your coins.

1.

 _____ ¢

2. _____ ¢

3. Q Q D N N N N _____ ¢

4. Q Q Q D D P P _____ ¢

5. Q Q Q Q Q D N _____ ¢

Date _____

1. Are you more likely to spin black or white?

_____

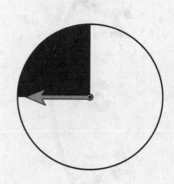

2. Measure your calculator to the nearest centimeter.

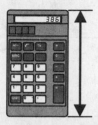

It measures about

_____ cm.

3. Write the fact family.

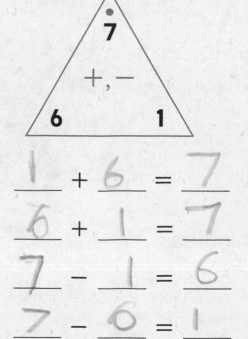

7

+, −

6   1

$1 + 6 = 7$

$6 + 1 = 7$

$7 - 1 = 6$

$7 - 6 = 1$

4. Shade all of the circles.

Date _____

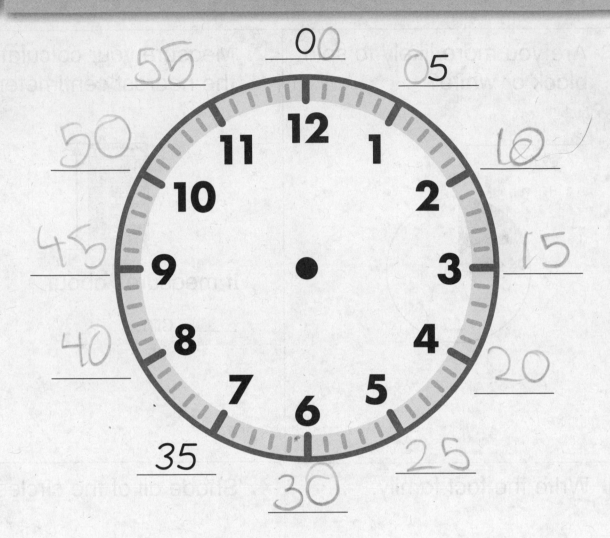

How many minutes are there in:

1. 1 hour? _____ minutes

2. Half an hour? __30__ minutes

3. A quarter-hour? _____ minutes

4. Three-quarters of an hour? _____ minutes

LESSON 6·10 Digital Notation

Draw the hour hand and the minute hand.

1.

4:00

2.

2:30

3.

6:15

Write the time.

4.

____:____

5.

____:____

6.

____:____

Make up your own. Draw the hour hand and minute hand. Write the time.

7.

____:____

8.

____:____

9.

____:____

LESSON 6·10 **Math Boxes**

1. Measure your shoe.

It measures about

_____ cm.

2. How much money?

_____ ¢

Use Ⓟ, Ⓝ, Ⓓ, and Ⓠ to show this amount with fewer coins.

3. Write <, >, or =.

ⓃⓃ ▢ Ⓓ

20¢ ▢ ⓃⓅ

24¢ ▢ $0.18

ⓄⒹⒹ ▢ 40¢

4. Count up by 10s.

50, _____, _____,

_____, _____, _____,

_____, _____, _____

LESSON 6·11   *My Reference Book* Scavenger Hunt

START
Turn to the Table of Contents.

Which section is about Measurement?
Fill in the circle next to the best answer.

(A) 5th section   (B) 3rd section

(C) 4th section   (D) 6th section

Find 2 tools in the measurement section. Draw them.

This is on page _____.   This is on page _____.

Turn to page 96.

Draw one pattern that you see.

Turn to the Games Section.

Find your favorite first-grade math game.

My favorite first-grade math game is

_____.

My favorite first-grade math game is on page

_____.

END

LESSON 6·11 **1- and 10-Centimeter Objects**

1. Find 3 things that are about 1 centimeter long.

Use words or pictures to show the things you found.

---

2. Find 3 things that are about 10 centimeters long.

Use words or pictures to show the things you found.

LESSON 6·11 Math Boxes

1. Are you more likely to spin black or white?

_____

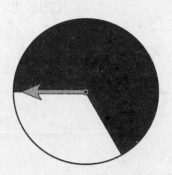

MRB 47 and 48

2. How many centimeters long is the line?

_____

Fill in the circle next to the best answer.

Ⓐ about 2 cm

Ⓑ about 3 cm

Ⓒ about 6 cm

Ⓓ about 7 cm

MRB 66

3. Write the fact family.

Total	
9	
Part	Part
9	0

_____ + _____ = _____

_____ + _____ = _____

_____ − _____ = _____

_____ − _____ = _____

MRB 25–27

4. Shade all of the squares.

MRB 55

LESSON 6·12 Class Results of Calculator Counts

1. I counted to _____ in 15 seconds.

2. Class results:

Largest count	Smallest count	Range of class counts	Middle value of class counts
_____	_____	_____	_____

3. Make a bar graph of the results.

Results of Calculator Counts

Number of Children

Counted to

Date _____

1. Draw a line segment that is about 7 centimeters long.

 (MRB 66)

2. How much money is ⓆⓆⓆⓅⓅⓅ?

Fill in the circle next to the best answer.

Ⓐ 78¢

Ⓑ 33¢

Ⓒ 73¢

Ⓓ 45¢

 (MRB 88 and 89)

3. Write <, >, or =.

7 + 6 ☐ 12

13 ☐ 6 + 7

14 − 6 ☐ 7

8 ☐ 15 − 6

 (MRB 9)

4. Count up by 5s.

25, _____, _____,

_____, _____, _____,

_____, _____, _____

LESSON 6·13 Math Boxes

1. Draw lines to match the shapes that look alike.

Column A   Column B

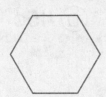

2. Shade the biggest square.

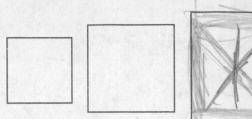

3. Draw lines to match the shapes that look alike.

Column A   Column B

4. Shade all of the triangles.

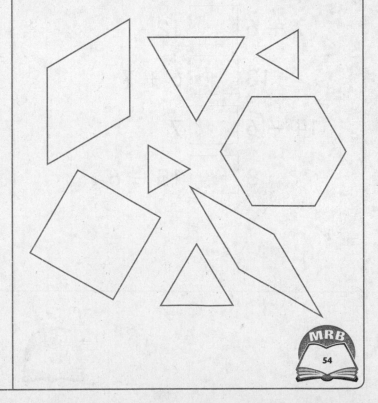

LESSON 7·1 *Make My Design*

Materials ☐ pattern blocks
☐ folder

Players 2

Skill Create designs using pattern blocks

Object of the Game To create a design identical to the other player's design

Directions:

1. The first player chooses 6 blocks. The second player gathers the same 6 blocks.

2. Players sit face-to-face with a folder between them.

3. The first player creates a design with the blocks.

4. Using only words, the first player tells the second player how to "Make My Design." The second player can ask questions about the instructions.

5. Players remove the folder and look at the two designs. Players discuss how closely the designs match.

6. Players change roles and play again.

LESSON 7·1 **Math Boxes**

1. Shade the large shapes.

2. Find the sums.
Circle the even sums.

$$6 + 8 \qquad 9 + 4$$

$$\_\_\_\_ = 8 + 9$$

$$5 + 7 = \_\_\_\_$$

96–97

3. Draw and solve.

There are 6 birds on a fence.
4 birds fly away.

How many birds are left?

_____ birds

4. Show 53¢ in two ways.

Use ⓠ, ⓓ, ⓝ, and ⓟ.

88–89

LESSON 7·2   **Math Boxes**

---

1. Draw what comes next.

☐   ☐☐   ☐☐☐   ☐☐☐☐   _____

---

2. Subtract.

$5 - 1 =$ _____   _____ $= 4 - 0$

$$\begin{array}{r} 3 \\ -\ 3 \\ \hline \end{array}$$ $$\begin{array}{r} 6 \\ -\ 1 \\ \hline \end{array}$$

---

3. Draw the hands.

2:30

4. Complete this part of the number grid.

	82	83
91	92	93
101	102	103
111	112	113
121	122	123

Date

Pattern-Block Template Shapes

1. Use your template to draw each shape.

square	large triangle	small hexagon
trapezoid	**small triangle**	**fat rhombus**
large circle	**skinny rhombus**	**large hexagon**

LESSON 7·3 — Pattern-Block Template Shapes *continued*

2. Draw shapes that have exactly 4 sides and 4 corners. Write their names.

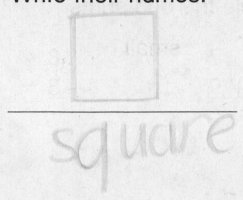

square

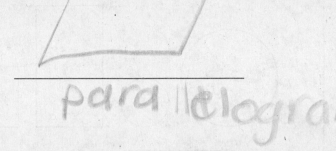

parallelogram

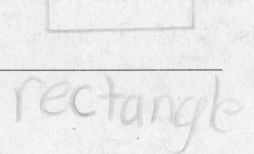

rectangle

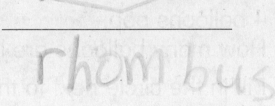

rhombus

LESSON 7·3 Math Boxes

1. Find the small square.
   Shade it.

2. Find the sums.
   Circle the odd sums.

$$7 + 9$$ $$8 + 3$$

$$6 + 7 = \underline{\hspace{1cm}}$$

$$\underline{\hspace{1cm}} = 10 + 4$$

3. Draw and solve.

   There are 8 balloons.
   4 balloons pop.
   How many balloons are left?

   Fill in the circle next to the best answer.

   ○ **A.** 0   ○ **B.** 6

   ○ **C.** 4   ○ **D.** 12

4. Show 81¢ in two ways.

   Use Ⓠ, Ⓓ, Ⓝ, and Ⓟ.

LESSON 7·4 **Polygons**

Triangles

4-Sided Polygons

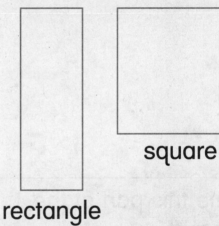

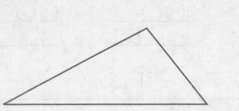

quadrangle

square

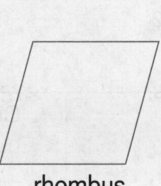

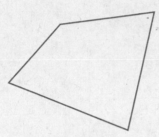

rectangle

rhombus   trapezoid

Some Other Polygons

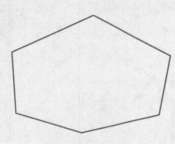

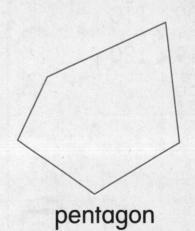

hexagon

pentagon

concave
hexagon

LESSON 7·4 **Math Boxes**

1. Draw what comes next.

2. Subtract.

$$\begin{array}{r} 2 \\ -1 \\ \hline 1 \end{array} \qquad \begin{array}{r} 4 \\ -4 \\ \hline 0 \end{array}$$

$$\begin{array}{r} 3 \\ -2 \\ \hline 1 \end{array} \qquad \begin{array}{r} 6 \\ -0 \\ \hline 6 \end{array}$$

3. Record the time.

8 : 15

4. Complete this part of the number grid.

88	89	90
98	99	100
108	109	110
118	119	120
128	129	130

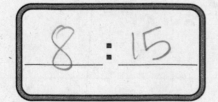

LESSON 7·5 | **Math Boxes**

1. Circle the name of this 3-dimensional shape.

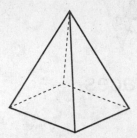

pyramid   prism

MRB 56

2. Circle the 4 polygons.

MRB 52–53

3. First-Grade Heights

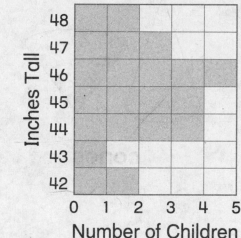

Inches Tall / Number of Children

Least inches tall: 42 inches

Most inches tall: 46 inches

 MRB 44

4. Draw a line to cut the pizza in half.

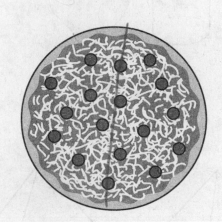

LESSON 7·6 Pictures of 3-Dimensional Shapes

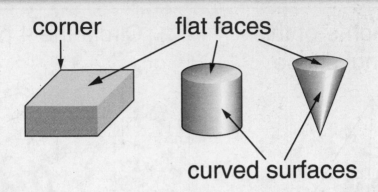

corner   flat faces

curved surfaces

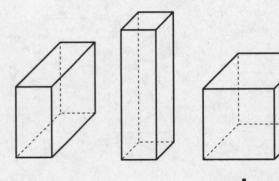

cube

rectangular prisms

sphere

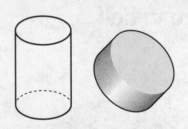

cylinders

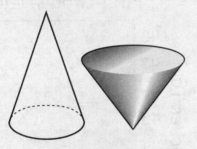

cones

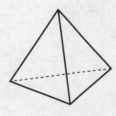
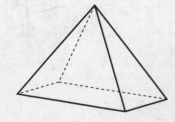
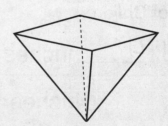
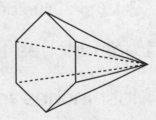

pyramids

Date _____

Identifying 3-Dimensional Shapes

What kind of shape is each object?
Write its name under the picture.

1.

_____

2.

_____

3.

_____

4.

_____

5.

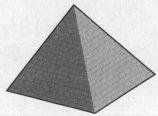

_____

6.

_____

LESSON 7·6 Math Boxes

1. Draw what comes next.

2. Subtract.

$$\begin{array}{r} 7 \\ -\ 0 \\ \hline \end{array}$$

$5 - 5 = \underline{\hspace{2cm}}$

$$\begin{array}{r} 5 \\ -3 \\ \hline \end{array}$$

$$\begin{array}{r} 4 \\ -2 \\ \hline \end{array}$$

3. What time is it?

Fill in the circle next to the best answer.

○ **A.** 9:30   ○ **B.** 6:09

○ **C.** 7:45   ○ **D.** 6:45

4. Complete this part of the number grid.

104	105	106
	115	116
124	125	
		136
144	145	

LESSON 7·7 **Math Boxes**

1. Name or draw 3 cylinders in your classroom.

MRB 57

2. Name this shape.

Fill in the circle next to the best answer.

○ **A.** rhombus

○ **B.** trapezoid

○ **C.** hexagon

○ **D.** square

MRB 54–55

3. **First-Grade Heights**

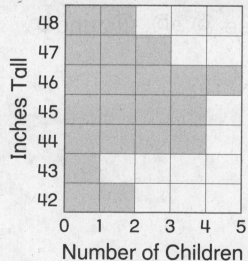

Inches Tall

Number of Children

What is the middle value?

About _____ inches

MRB 46

4. Draw a line to cut the cookie in half.

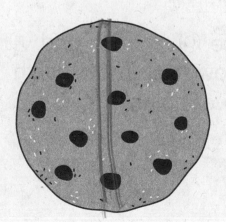

Math Boxes

1. Divide each shape in half.

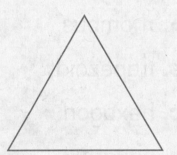

2. Complete this part of the number grid.

123	124	125
133	134	135
143	144	145
153	154	155
163	164	165

3. How much money?

Ⓠ Ⓝ Ⓓ Ⓝ Ⓓ Ⓟ Ⓝ

_____ ¢

Use Ⓠ, Ⓓ, Ⓝ, and Ⓟ to show this amount with fewer coins.

4. Show 65¢ in two ways.

Use Ⓠ, Ⓓ, Ⓝ, and Ⓟ.

How Much Money?

Record the amount shown.

1.

2.

_____ ¢   _____ ¢

Mark the coins you need to buy each item.

3.

86¢
crystal

4.

59¢
horse

LESSON 8·1 **Time**

Draw the hands.

1.

10:00

2.

6:30

3.

1:45

4.

8:15

5.

11:05

6.

2:35

LESSON 8·1 Math Boxes

1. How much money?

_____ 134 ¢ or

$ _____

2. Draw a line to match each face to the correct picture of the 3-dimensional shape.

Column A   Column B

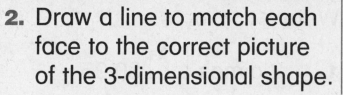

MRB
88–90

MRB
58

3. Solve.

8	4
+ 7	+ 9

14	17
− 6	− 9

4. Use your number grid.

Start at 59.

Count back 20.

_____ = 59 − 20

MRB
32

LESSON 8·2 Comparing Money Amounts

Write <, >, or =.

> < is less than
> = is equal to
> > is greater than

1. 2 dimes ☐ $0.25

2. 50¢ ☐ 5 pennies

3. 4 quarters ☐ 100¢

4. 100¢ ☐ 20 nickels

5. $1.25 ☐ 120¢

6. $1.75 ☐ 7 quarters

7. 200¢ ☐ 10 dimes and 10 nickels

8. $1.44 ☐ 1 dollar, 4 dimes, and 4 pennies

LESSON 8·2

Math Boxes

1. Use , and 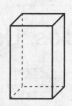 to show this amount.

$3.49

MRB
88–90

2. This is a picture of a shape. Circle the name of the shape.

prism   pyramid

MRB
56–57

3. Subtract.

18	12
− 9	− 8

15	10
− 7	− 7

4. Add.

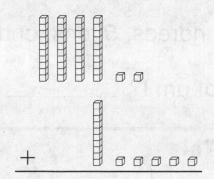

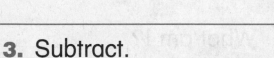

MRB
28

LESSON 8·3 Hundreds, Tens, and Ones Riddles

Hundreds	Tens	Ones

Solve the riddles. Use your base-10 blocks to help you.

Example: 2 ▦ 3 ▯ 5 ▱

What am I? __235__

1. 7 ▯ 2 ▱

What am I? __72__

2. 2 ▦ 3 ▯ 4 ▱

What am I? __234__

3. 8 hundreds, 5 tens, and 2 ones

What am I? __852__

4. 4 hundreds and 6 ones

What am I? __406__

Try This

5. 2 hundreds, 14 tens, and 5 ones. What am I? _____

6. 12 ones, 7 tens, and 3 hundreds. What am I? _____

7. Make up your own riddle. Ask a friend to solve it.

LESSON 8·3

Math Boxes

1. Count the coins.

Choose the best answer.

- Ⓐ 38¢
- Ⓑ 88¢
- Ⓒ 93¢
- Ⓓ 83¢

MRB 88–90

2. Draw a line to match each face to the correct picture of the 3-dimensional shape.

Column A   Column B

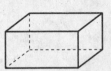

MRB 58

3. Solve.

$$3 + 7$$ $$7 + 9$$

$$16 - 6$$ $$14 - 9$$

4. Use your number grid.

Start at 71.

Count up 19.

$$71 + 19 = \underline{\hspace{1cm}}$$

MRB 29

LESSON 8·4 School Store Mini-Poster 2

crayon
6¢

scissors
32¢

ball
35¢

gum
2¢

pencil
28¢

candy
8¢

eraser
17¢

Date _____

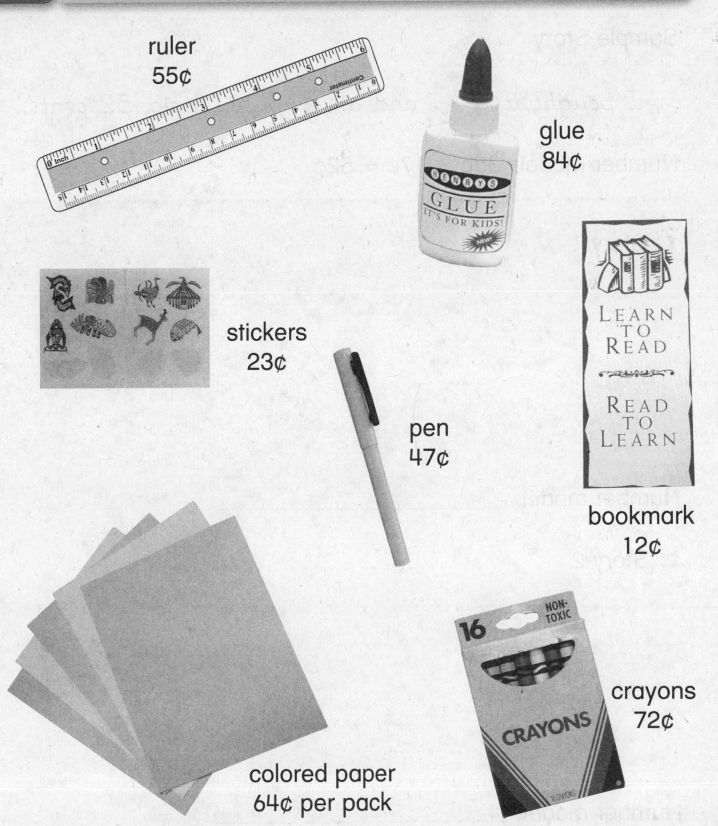

ruler
55¢

glue
84¢

stickers
23¢

pen
47¢

bookmark
12¢

colored paper
64¢ per pack

crayons
72¢

Number Stories

Sample Story

I bought a and an . I paid 52 cents.

Number model: 35¢ + 17¢ = 52¢

1. Story 1

Number model: _____

2. Story 2

Number model: _____

LESSON 8·4 Math Boxes

1. $1.00 =

_____ pennies

_____ nickels

_____ dimes

_____ quarters

MRB 88–90

2. Name or draw 2 objects shaped like a rectangular prism.

MRB 56–57

3. Subtract.

13	11
− 6	− 9

16	14
− 8	− 8

2. What is the sum?

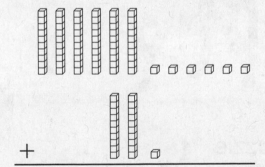

Choose the best answer.

- Ⓐ 45
- Ⓑ 87
- Ⓒ 85
- Ⓓ 78

MRB 28

LESSON 8·5 Museum Store Mini-Poster

seashell
48¢

kite
$1.86

elephant
72¢

rock
35¢

horse
59¢

ring
18¢

magnet
$1.39

puzzle
85¢

plane
27¢

LESSON 8·5 Making Change

Record what you bought. Record how much change you got.

Example:

I bought _____*a plane*_____ for ___27___ cents.

I gave _____Ⓓ Ⓓ Ⓓ_____
to the clerk.

I got _____Ⓟ Ⓟ Ⓟ_____ in change.

_____

1. I bought _____ for _____ cents.

I gave _____ to the clerk.

I got _____ in change.

_____

2. I bought _____ for _____ cents.

I gave _____ to the clerk.

I got _____ in change.

_____

3. I bought _____ for _____ cents.

I gave _____ to the clerk.

I got _____ in change.

LESSON 8·5

Math Boxes

1. Circle the tens place.

36   120   59

66   20   104

MRB 10

2. Keisha bought a ball for 25¢.

She bought a bat for 75¢.

How much did Keisha pay?

_____ ¢ or $_____ . _____

3. Draw the other half.

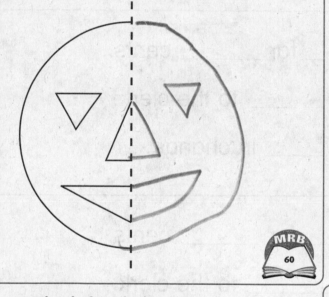

MRB 60

4. Draw a polygon with 6 sides.

MRB 52–53

5. Label the box. Add 3 new names.

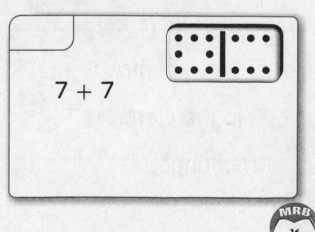

7 + 7

MRB 16

6. Subtract.

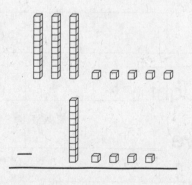

MRB 31

Date _____

Equal Shares

Show how you share your crackers.

1 cracker, 2 people

Halves

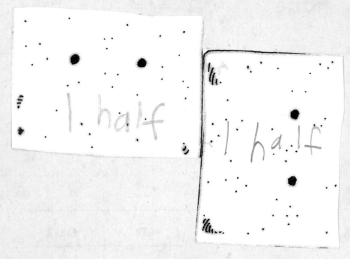

1 cracker, 4 people

Fourths

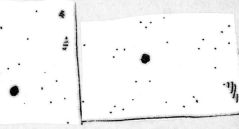

1 cracker, 3 people

Thirds

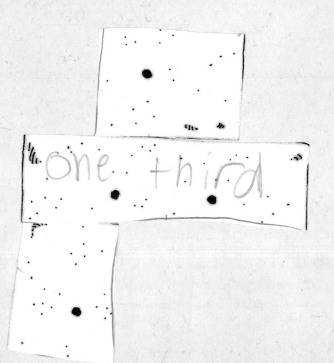

2 crackers, 4 people

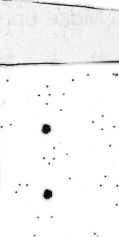

LESSON 8·6 "What's My Rule?"

Write the rule. Complete the table.

1.

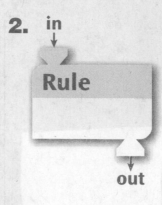

in → Rule → out

in	out
23	33
15	25
7	
37	

2.

in → Rule → out

in	out
13	3
51	41
	8
29	

3.

in → Rule → out

in	out
45	25
21	1
70	
	37

4.

in → Rule → out

in	out
12	32
28	48
30	
	65

Make up your own.

5.

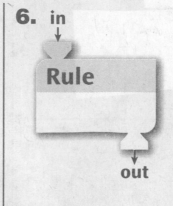

in → Rule → out

in	out

6.

in → Rule → out

in	out

LESSON 8·6

Math Boxes

1. A ring costs 20¢.

I pay a Ⓠ.

How much change will
I get?

_____ ¢

2. How many equal parts?

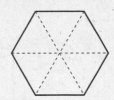

Choose the best answer.

Ⓐ 8   Ⓑ 1   Ⓒ 6   Ⓓ 2

MRB 12

3. What number are you most likely to spin?

_____

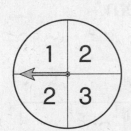

MRB 47–48

4. Write the fact family.

____ + ____ = ____

____ + ____ = ____

____ − ____ = ____

____ − ____ = ____

MRB 25

5. Complete the graph.

5 children take the bus.
3 children ride bikes.

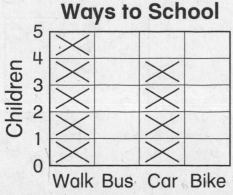

MRB 44

6. Use <, >, or =.

MRB 9, 11

LESSON 8·7 Equal Parts of Wholes

Which glass is half full? Circle it.

Which rectangles are divided into thirds? Circle them.

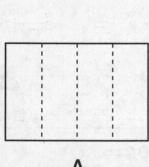

A

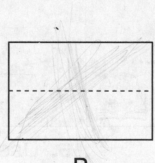

B

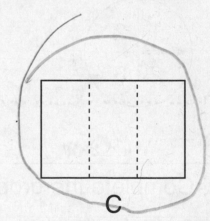

C

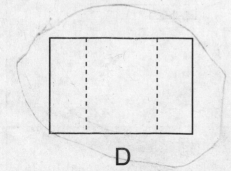

D

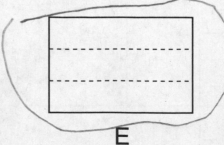

E

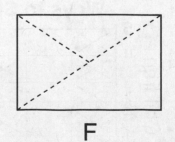

F

LESSON 8·7   Fractions

1. How many equal parts are there? __3__

   Write a fraction in each part of the circle.

   Color $\frac{1}{3}$ of the circle.

one third

2. How many equal parts are there? __4__

   Write a fraction in each part of the square.

   Color $\frac{1}{4}$ of the square.

one fourth

3. How many equal parts are there? __6__

   Write a fraction in each part of the hexagon.

   Color $\frac{1}{6}$ of the hexagon.

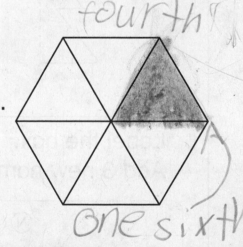

one sixth

4. How many equal parts are there? __8__

   Write a fraction in each part of the rectangle.

   Color $\frac{1}{8}$ of the rectangle.

One eigth

Date _____

1. Circle the hundreds place.

289   (300)   112

733   999   205

MRB 10

2. Carlos bought 3 pencils.
Each pencil costs 10¢.
How much did Carlos pay?

___30___ ¢ or $_____

3. Draw the other half.

MRB 60

4. Use a straightedge.

Draw line segments to make a polygon.

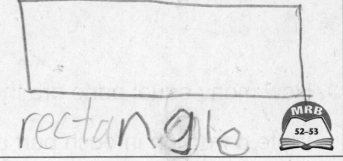

rectangle

MRB 52–53

5. Label the box.
Add 3 new names.

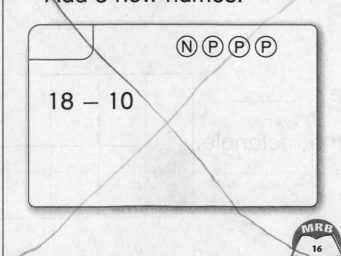

18 − 10

MRB 16

6. Subtract.

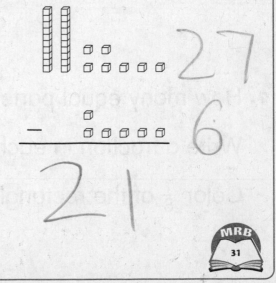

MRB 31

LESSON 8·8   Sharing Pennies

Use your pennies to help you solve the problems.

Circle each person's share.

1. Halves: 2 people share 8 pennies equally.

How many pennies does each person get? ____ pennies

2. Thirds: 3 people share 9 pennies equally.

How many pennies does each person get? ____ pennies

How many pennies do 2 of the 3 people get in all?

____ pennies

LESSON 8·8 **Sharing Pennies** *continued*

3. Fifths: 5 people share 15 pennies equally.

How many pennies does each person get? _____ pennies

How many pennies do 3 of the 5 people get in all?

_____ pennies

4. Fourths: 4 people share 20 pennies.

How many pennies does each person get? _____ pennies

How many pennies do 2 of the 4 people get in all?

_____ pennies

LESSON 8·8

Math Boxes

1. A seashell costs $0.48. I pay 2 Ⓠ.

How much change will I get?

______¢

2. Label each equal part.

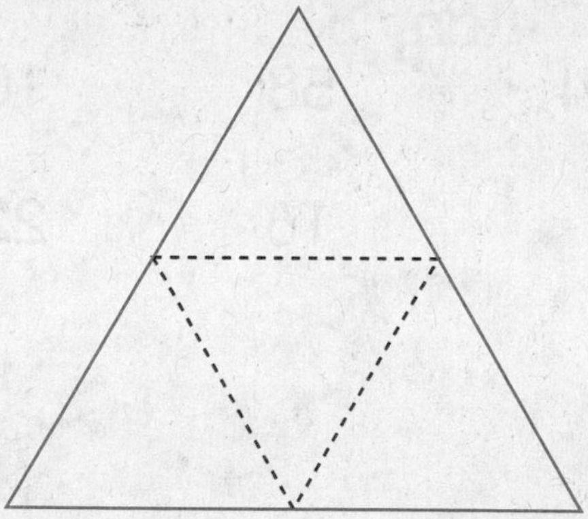

MRB 12–13

3. Complete the graph.

5 children live 6 blocks away.
4 children live 7 blocks away.

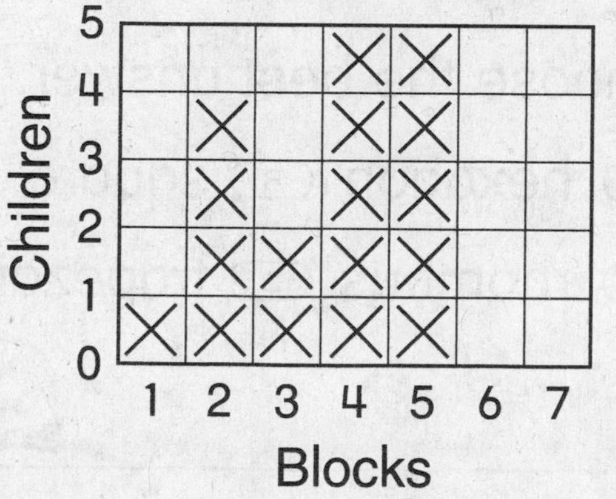
Blocks from School
Children
0
1
2
3
4
5
1
2
3
4
5
6
7
Blocks

MRB 44

4. Write the fact family.

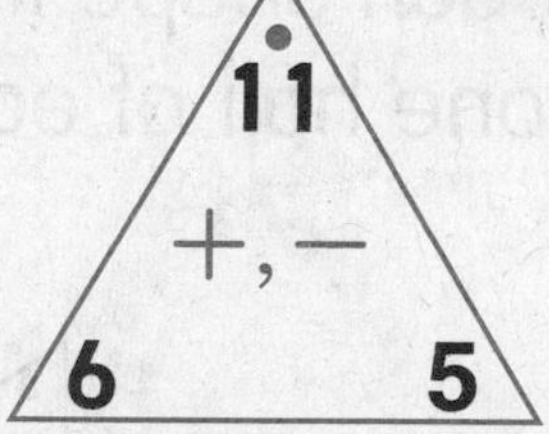
11
+, −
6
5

______ + ______ = ______

______ + ______ = ______

______ − ______ = ______

______ − ______ = ______

MRB 27

5. What number are you most likely to spin?

______

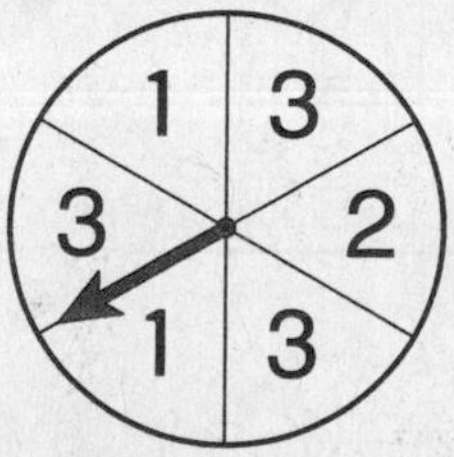
1
3
2
3
1
3

MRB 47–48

6. Write <, >, or =.

13 ☐ 31

108 ☐ 80

1 + 2 ☐ 12

MRB 9

LESSON 8·9 Math Boxes

1. Circle the ones place.

364   58   100

4   16   222

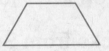

2. You buy 2 packs of seeds. Each pack costs 60¢.

How much do you pay?

_____ ¢ or $_____ .

3. Divide each shape in half. Shade one half of each shape.

4. Name this polygon.

Choose the best answer.

Ⓐ hexagon   Ⓑ square

Ⓒ rhombus   Ⓓ trapezoid

5. Write 3 more names.

100

80 + 20

6. Subtract.

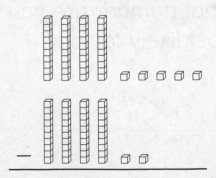

LESSON 8·10 — Math Boxes

1. Use <, >, or =.

Q Q ☐ $0.25

$1.00 ☐ Q D Q D

10¢ + 20¢ ☐ N Q

MRB 9, 88–90

2. Use your number grid.

Start at 33.
Count back 9.

33
− 9

MRB 32

3. Subtract.

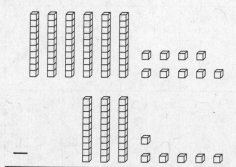

MRB 31

4. Add.

MRB 28

5. Solve.

$8 + 9 =$ _____

_____ $= 6 + 8$

_____ $= 19 - 9$

$13 - 8 =$ _____

6. Write 3 names.

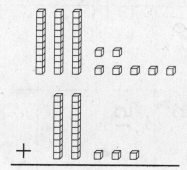

$1.00

MRB 16

LESSON 9·1 Number-Grid Hunt

-9	-8	-7	-6	-5	-4	-3	-2	-1	0
1	2	3	4	5	6	7	8	9	10
11	12	13	14	15	16	17	18	19	20
21	22	23	24	25	26	27	28	29	30
31	32	33	34	35	36	37	38	39	40
41	42	43	44	45	46	47	48	49	50
51	52	53	54	55	56	57	58	59	60
61	62	63	64	65	66	67	68	69	70
71	72	73	74	75	76	77	78	79	80
81	82	83	84	85	86	87	88	89	90
91	92	93	94	95	96	97	98	99	100
101	102	103	104	105	106	107	108	109	110

Date _____

1. Use your number grid.
Start at 26.
Count up 14.

$$26$$
$$+\ 14$$

2. Shade $\frac{1}{4}$ of the circle.

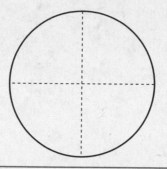

3. Fill in the missing numbers.

in →

Rule
Add 5

out ↓

in	out
8	
12	
29	
41	
100	

MRB
100–102

4. Write the fact family.

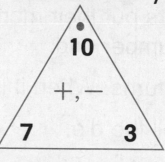

10

+, −

7   3

___ + ___ = ___

___ + ___ = ___

___ − ___ = ___

___ − ___ = ___

MRB
27

5. Tell the time.

____ : ____

MRB
81

6. Freddy has Ⓠ Ⓓ Ⓓ Ⓝ.
Jewel has Ⓓ Ⓓ Ⓝ Ⓓ Ⓓ Ⓟ.

Who has more money?

_____

How much more money?

_____ ¢

LESSON 9·2 Number-Grid Game

Materials	☐ a number grid
	☐ a die
	☐ a game marker for each player
Players	2 or more
Skill	Counting on the number grid
Object of the Game	To land on 110 with an exact roll

1. Players put their markers at 0 on the number grid.

2. Take turns. When it is your turn:
   ◆ Roll the die.
   ◆ Use the table to see how many spaces to move your marker.
   ◆ Move your marker that many spaces.

3. Continue playing. The winner is the first player to get to 110 with an exact roll.

Roll	Spaces
⚀	1 or 10
⚁	2 or 20
⚂	3
⚃	4
⚄	5
⚅	6

Date _____

1. Use your number grid.
   Start at 90.
   Count back 25.

   90
   − 25
   _____

   MRB
   32

2. Find the sums.

   2 + 5 = _____

   12 + 5 = _____

   42 + 5 = _____

   102 + 5 = _____

3. Shade $\frac{1}{2}$ of the pennies.

   Ⓟ Ⓟ Ⓟ Ⓟ Ⓟ

   Ⓟ Ⓟ Ⓟ Ⓟ Ⓟ

   MRB
   14

4. Asha bought a key chain
   for 43¢.

   She paid Ⓠ Ⓠ.

   How much change did
   she get? _____ ¢

   Show this amount with Ⓓ,
   Ⓝ, and Ⓟ.

5. Fill in the missing numbers.

   MRB
   98, 99

   Rule

   −1

653	652				

6. Circle the four polygons.

   MRB
   52, 53

LESSON 9·3

Number-Grid Puzzles 1

0	10	20	30	40		60	70	80	90	100	110
	9										
						68	78		79		
	7	17	27	37		67					
						66			95		
	5								85		
	4			44	54	64	74	84		104	
	3	23						83		103	
	2	22	32	42	52		72			102	
		21				61			91	101	

LESSON 9·3 **Math Boxes**

---

1. Use your number grid.
Start at 36.
Count up 22.

36 + 22 = _____

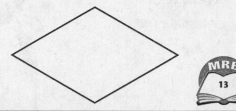
MRB 29

2. Divide the rhombus in half.
Shade $\frac{1}{2}$.

MRB 13

---

3. Find the rule and the missing numbers.

in	out
15	5
21	
84	74
	94

Rule

MRB 100–102

4. Write the fact family.

_____ + _____ = _____

_____ + _____ = _____

_____ − _____ = _____

_____ − _____ = _____

MRB 27

---

5. What time is it?

Fill in the circle next to the best answer.

Ⓐ 2:20

Ⓑ 4:02

Ⓒ 4:10

Ⓓ 2:04

MRB 81

6. Jonah has Ⓠ Ⓓ Ⓓ Ⓠ Ⓟ Ⓟ Ⓟ.

Mari has Ⓠ Ⓓ Ⓝ Ⓝ Ⓓ Ⓠ Ⓟ.

Who has more money?

_____

How much more money?

_____¢

Date _____

Silly Animal Stories

Example:

Unit
inches

koala
24 in.

penguin
36 in.

How tall are the koala
and penguin together?

24 + 36 = 60

60 *inches*

---

1. Silly Story

Unit

---

2. Silly Story

Unit

Date

1. Use your number grid. Start at 48. Count back 15.

48 − 15 = ?

Fill in the circle next to the best answer.

Ⓐ 43　　Ⓑ 33

Ⓒ 63　　Ⓓ 36

MRB 32

2. Solve.

16 − 9 = ____

26 − 9 = ____

56 − 9 = ____

106 − 9 = ____

3. Draw 12 dimes. Use Ⓓs.

Shade $\frac{1}{4}$ of the dimes.

MRB 14

4. A toy dinosaur costs 89¢.

I paid $1.00.

How much change do I get?

_____¢

Show this amount with Ⓓ, Ⓝ, and Ⓟ.

5. Find the rule. Fill in the missing numbers.

MRB 98, 99

Rule

| 165 | 265 | 365 | | |

6. Circle the 3 polygons.

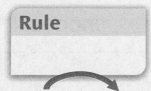

MRB 52, 53

one hundred eighty-three **183**

LESSON 9·5 My Height Record

First Measurement

Date _____

Height: about _____ inches

A typical height for a first grader in my class was about

_____ inches.

Second Measurement

Date _____

Height: about _____ inches

A typical height for a first grader in my class is about

_____ inches.

The middle height for my class is about _____ inches.

Change to Height

I grew about _____ inches.

The typical growth in my class was about _____ inches.

LESSON 9·5 Math Boxes

1. Find the sums.

2 + 6 = _____

20 + 60 = _____

200 + 600 = _____

2. Complete the number-grid puzzle.

42

MRB
7

3. Label each part.

Shade $\frac{1}{3}$ of the rectangle.

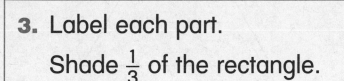

MRB
13

4. Circle the 4 letters that are symmetrical.

A P H E

L V F Q

MRB
60

5. Myla buys 2 items at the store.

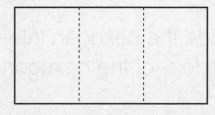

45¢

TOOTHPASTE

39¢

How much money does she spend?

_____

Show this amount with
Ⓠ, Ⓓ, Ⓝ, and Ⓟ.

6. This is a picture of a 3-D shape.
Circle the name of the shape.

pyramid   cone

MRB
56, 57

LESSON 9·6 Pattern-Block Fractions

Use pattern blocks to divide each shape into equal parts.
Draw the parts using your Pattern-Block Template.
Shade parts of the shapes.

1. Divide the rhombus into halves. Shade $\frac{1}{2}$ of the rhombus.

2. Divide the trapezoid into thirds. Shade $\frac{2}{3}$ of the trapezoid.

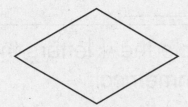

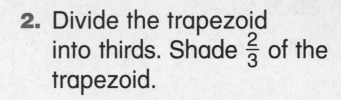

3. Divide the hexagon into halves. Shade $\frac{2}{2}$ of the hexagon.

4. Divide the hexagon into thirds. Shade $\frac{2}{3}$ of the hexagon.

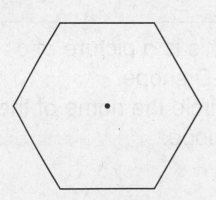

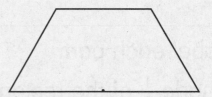

5. Divide the hexagon into sixths. Shade $\frac{4}{6}$ of the hexagon.

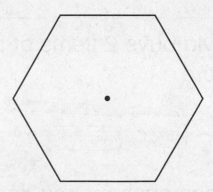

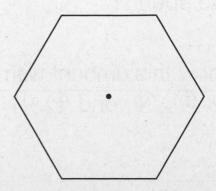

1. Solve.

_____ = 9 + 9

_____ = 90 + 90

$$\begin{array}{ccc} 7 & 70 & 700 \\ -5 & -50 & -500 \end{array}$$

2. Divide the rectangle into fourths. Shade $\frac{3}{4}$ of the rectangle.

MRB 12, 13

3. Draw and solve.

Griffin had 14 guppies.

He gave $\frac{1}{2}$ away.

How many guppies are left?

_____ guppies

MRB 14

4. Write the numbers.

_____

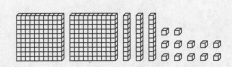

_____

MRB 10, 11

5. **Weekly Allowance**

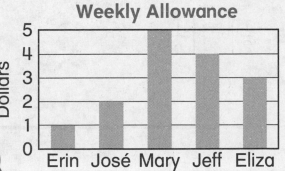

MRB 44

Smallest allowance: $ _____

Largest allowance: $ _____

6. Record the temperature.

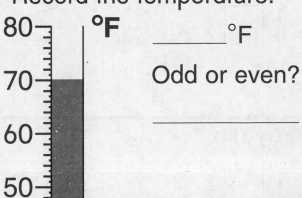

_____ °F

Odd or even?

_____

MRB 87, 97

LESSON 9·7 Fraction Strips

Use your fraction strips to compare the fractions.

> < is less than
> > is greater than
> = is equal to

Unit
1-strip

1. $\frac{1}{2}$ ☐ $\frac{1}{4}$

2. $\frac{1}{8}$ ☐ $\frac{1}{4}$

3. $\frac{1}{2}$ ☐ $\frac{1}{8}$

4. $\frac{1}{2}$ ☐ $\frac{1}{3}$

5. $\frac{1}{6}$ ☐ $\frac{1}{3}$

6. $\frac{1}{4}$ ☐ $\frac{1}{3}$

7. $\frac{1}{4}$ ☐ $\frac{1}{6}$

8. $\frac{1}{2}$ ☐ $\frac{1}{6}$

Try This

9. $\frac{1}{2}$ ☐ $\frac{2}{3}$

10. $\frac{2}{4}$ ☐ $\frac{1}{2}$

Date _____

1. Solve.

7 − 4 = _____

70 − 40 = _____

700 − 400 = _____

2. Complete the number-grid puzzle.

| 79 | |

MRB 7

3. Label each part. Shade $\frac{5}{6}$ of the hexagon.

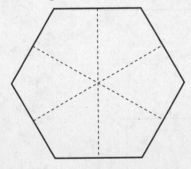

MRB 13

4. Circle the 3 numbers that are symmetrical.

1 6 3

0 5 4

MRB 60

5. Diego buys 2 items at the store.

$1.50 CRAYONS
DRAWING PAPER 29¢

How much money does he spend?
$_____
Show this amount with $1,
 Q, D, N, and P.

6. This is a picture of a 3-dimensional shape. Name the shape.

Fill in the circle next to the best answer.

Ⓐ sphere   Ⓑ cube

Ⓒ cylinder   Ⓓ cone

MRB 56, 57

LESSON 9·8 Many Names for Fractions

1-strip

Use your fraction pieces to help you solve the following problems.

Example:

4 | $\dfrac{1}{8}$ | = | $\dfrac{1}{2}$

$\dfrac{4}{8} = \dfrac{1}{2}$

1.

___ | $\dfrac{1}{6}$ | = | $\dfrac{1}{3}$

$\dfrac{}{6} = \dfrac{1}{3}$

2.

___ | $\dfrac{1}{8}$ | = | $\dfrac{1}{4}$

$\dfrac{}{8} = \dfrac{1}{4}$

LESSON 9·8

Many Names for Fractions *continued*

3.

$\dfrac{1}{6}$	$=$	$\dfrac{1}{3}$	$\dfrac{1}{3}$

$$\dfrac{}{6} = \dfrac{2}{3}$$

4.

$\dfrac{1}{8}$	$=$	$\dfrac{1}{4}$	$\dfrac{1}{4}$

$$\dfrac{}{8} = \dfrac{2}{4}$$

5.

$\dfrac{1}{8}$	$=$	$\dfrac{1}{4}$	$\dfrac{1}{4}$	$\dfrac{1}{4}$

$$\dfrac{}{8} = \dfrac{3}{4}$$

LESSON 9·8 **Math Boxes**

---

1. Solve.

_____ = 9 − 5

_____ = 90 − 50

$$\begin{array}{ccc} 6 & 60 & 600 \\ +\ 4 & +\ 40 & +\ 400 \\ \hline \end{array}$$

2. What fraction is shaded? Fill in the circle next to the best answer.

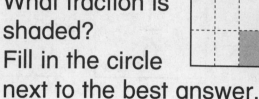

Ⓐ $\frac{1}{3}$ 　Ⓑ $\frac{3}{8}$

Ⓒ $\frac{8}{3}$ 　Ⓓ $\frac{3}{1}$

13

---

3. Draw and solve. Emma had 15 grapes. She gave $\frac{1}{3}$ to her sister. How many grapes did her sister get?

_____ grapes

14

4. Write the numbers.

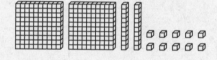

_____

_____

10, 11

---

5. Books Read in a Week

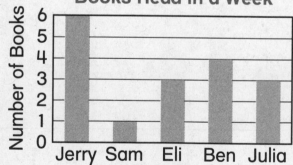

Jerry Sam Eli Ben Julia

Fewest number of books read: _____

Greatest number of books read: _____

Range: _____

44, 45

6. Record the temperature.

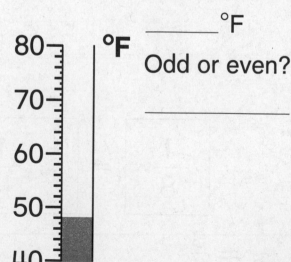

_____ °F

Odd or even?

_____

87, 97

LESSON 9·9 | **Math Boxes**

---

1. Lowest count:

_____

Highest count:

_____

Range:

_____

Calculator Counts in 15 Seconds

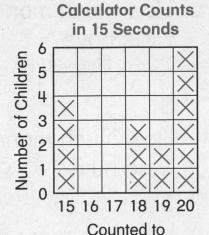

Number of Children (y-axis: 0–6)
Counted to (x-axis: 15 16 17 18 19 20)

MRB 44, 45

---

2. Sam buys 2 items at the store.

How much money does he spend?

_____

Show this amount with $1, Q, D, N, and P.

MRB 88–90

---

3. You have $1.00. You buy a pretzel that costs 75¢.

How much change do you get?

_____ ¢

Show this amount with Q, D, N, and P.

---

4. Pedro has Q D N P N N D. Claudia has D D N Q N Q P.

Who has more money?

_____

How much more money?

_____ ¢

---

5. Use a ruler. Draw a polygon with 4 sides.

MRB 52–53

---

6. Draw the hands.

9:35

MRB 81

LESSON 10·1 Math Boxes

1. Record the temperatures.

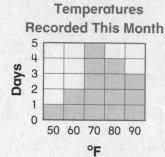

Temperatures
Recorded This Month

Coolest temperature:

_____°F

Warmest temperature: _____°F

Range: _____°F

MRB 44–45

2. How much money does Kylie have?

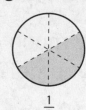

$_____.____

How much money does Pete have?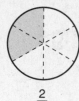

$_____.____

Who has more money?

_____

How much more money?

$_____.____

3. Draw a triangle with one side that is 1 inch long.

MRB 54

4. Circle the bigger fraction.

$\frac{2}{6}$ $\frac{1}{2}$

MRB 13

5. Fill in the missing numbers.

in
↓

Rule	in	out
+100	25	
	99	
	174	
	565	

out ↓

MRB 100–101

6. Draw a polygon with 4 sides.

MRB 52–53

Date _____

1. Ask a partner to show times on a tool-kit clock.
   Draw the hands on the clock. Write the times to match.

_____ : _____   _____ : _____   _____ : _____

2. Write a time for each clock face. Draw the hands to match.

_____ : _____   _____ : _____   _____ : _____

3. Set your tool-kit clock to 3:00.
   How many minutes until 3:25? _____ minutes

   Set your tool-kit clock to 1:30.
   How many minutes until 1:55? _____ minutes

   Set your tool-kit clock to 10:45.
   How many minutes until 11:20? _____ minutes

LESSON 10·2 Math Boxes

1. Draw and solve.

There are 8 cups.

5 cups are dirty.

How many cups are clean?

_____ cups

2. I buy a kite for $1.89.

I pay $2.00.

How much change do I get back?

_____ ¢

3. Write <, >, or =.

305 ☐ 385

113 ☐ 100 + 13

129 ☐

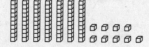

4. Complete the number-grid puzzle.

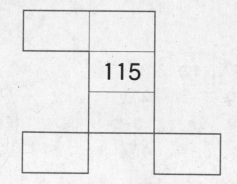

115

5. Write the number that is 10 more.

_____   _____

6. Fill in the rule and the missing numbers.

Rule

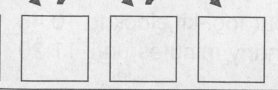

266   268   ☐   ☐   ☐

LESSON 10·3 Vending Machine Poster

LESSON 10·3 Buying from the Vending Machine

Pretend to buy items from the vending machine. Draw pictures or write the names of items you buy. Show the coins you use to pay for the items. Use Ⓝ, Ⓓ, and Ⓠ. Write the total cost.

1.

2.

3.

4.

Show the cost of these items. Use Ⓝ, Ⓓ, and Ⓠ.
Write the total cost.

5.

Total cost: $_____

6.

Total cost: $_____

1. Record the times.

First-Grade Bedtimes

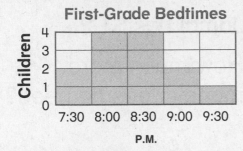

Earliest bedtime: _____

Latest bedtime: _____

Range: _____ hours

MRB 44–45

2. Clay has $1

Rosa has

Who has more money?

_____

How much more money?

_____ ¢

3. Measure the line segment.

It is about _____ inches long.

_____

Fill in the circle next to the best answer.

○ **A.** 9   ○ **B.** 3

○ **C.** $3\frac{1}{2}$ ○ **D.** $8\frac{1}{2}$

MRB 65

4. Circle the bigger fraction.

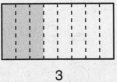

$\frac{3}{4}$ $\frac{3}{8}$

MRB 13

5. Fill in the missing numbers.

in →

Rule

+10

→ out

in	out
___	18
___	86
___	103
___	264

MRB 100–101

6.

How many sides?

____ sides

How many corners?

____ corners

MRB 52–53

Math Boxes

---

1. Draw and solve.

Amelia has 12 checkers.

6 checkers are black.

How many checkers are not black?

_____ checkers

---

2. A magnet costs $1.39.

Jamal has $1.25.

How much more money does he need?

_____¢

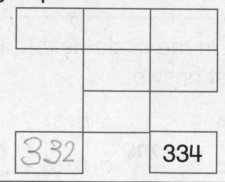

$1.39

---

3. Write <, >, or =.

Q Q Q Q Q ☐ $1.25

Q D D N D ☐ $0.50

Q Q Q D D D ☐ $1.00

MRB
9, 88, 89

---

4. Complete the number-grid puzzle.

332		334

---

5. Write the number that is 10 more.

 _____

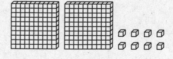 _____

MRB
10–11

---

6. Fill in the rule and the missing numbers.

Rule

 | | | 148 | 158 | |

MRB
98–99

Date _____

Triangles

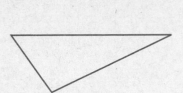

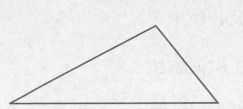

Quadrangles (Quadrilaterals)

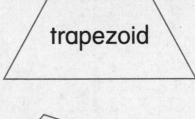
trapezoid

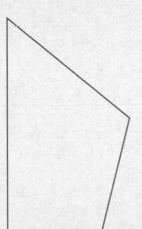

kite

rhombus

square

rectangle

Other Polygons

hexagon

octagon

pentagon

heptagon

LESSON 10·5   Reviewing Polygons

Use straws and twist-ties to make the following polygons.
Draw the polygons. Record the number of corners and sides
for each polygon.

1. Make a square.

   Number of sides _____

   Number of corners _____

2. Make a triangle.

   Number of sides _____

   Number of corners _____

3. Make a hexagon.

   Number of sides _____

   Number of corners _____

4. Make a polygon of your choice.

   Write its name. _____

   Number of sides _____

   Number of corners _____

5. Make another polygon of your choice.

   Write its name. _____

   Number of sides _____

   Number of corners _____

LESSON 10·5 Reviewing 3-Dimensional Shapes

Word Bank

sphere	rectangular prism	pyramid
cube	cone	cylinder

Write the name of each 3-dimensional shape.

1.

_____

2.

_____

3.

_____

4.

_____

5.

_____

6.

_____

Five Regular Polyhedrons

The faces that make each shape are identical.

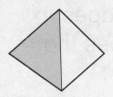 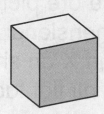 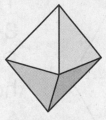 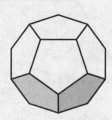 

tetrahedron	cube	octahedron	dodecahedron	icosahedron
4 faces	6 faces	8 faces	12 faces	20 faces

LESSON 10·5 | **Math Boxes**

---

1. Solve.

$\frac{1}{2}$ of 50¢ = _____ ¢

$\frac{1}{2}$ of $1.00 = _____ ¢

$\frac{1}{2}$ of $2.00 = $_____._____

2. Which are you more likely to grab?

black or white? _____

○ or □? _____

47–48

---

3. Record the time.

_____ : _____

80–81

4. Shade the thermometer to show 78° F.

87

---

5. Label each part.

Shade $\frac{2}{6}$ of the hexagon.

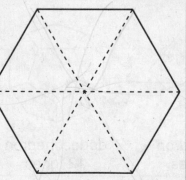

Write another name for $\frac{2}{6}$.

_____

13

6. These are pictures of 3-dimensional shapes. Put an X on shapes that have all flat faces.

pyramid   sphere   cylinder   cube

58

LESSON 10·6

U.S. Weather Map

U.S. Weather Map: Spring High/Low Temperatures (°F)

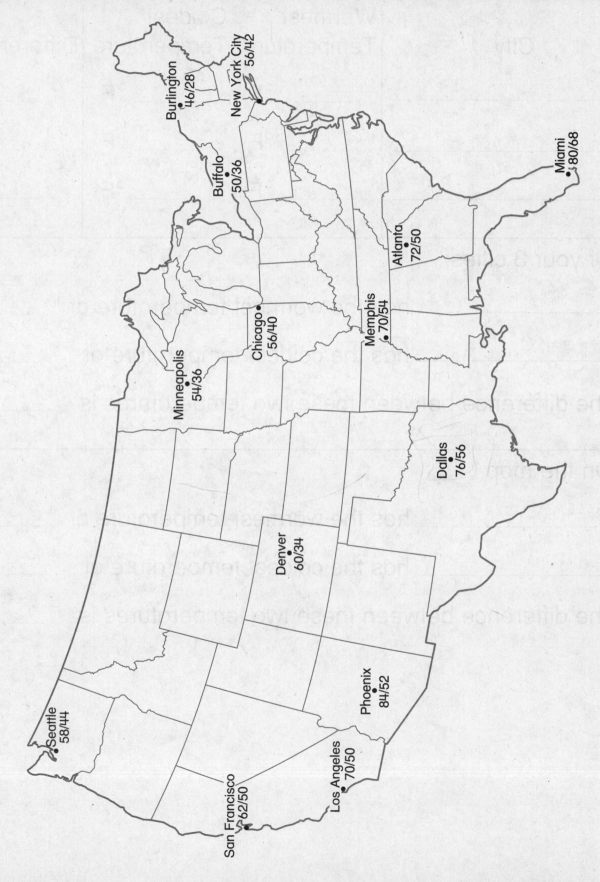

Burlington
46/28

New York City
56/42

Buffalo
50/36

Miami
80/68

Atlanta
72/50

Chicago
56/40

Memphis
70/54

Minneapolis
54/36

Dallas
76/56

Denver
60/34

Phoenix
84/52

Seattle
58/44

Los Angeles
70/50

San Francisco
62/50

Temperature Chart

1.

City	Warmest Temperature	Coldest Temperature	Difference
_____	_____ °F	_____ °F	_____ °F
_____	_____ °F	_____ °F	_____ °F
_____	_____ °F	_____ °F	_____ °F

2. Of your 3 cities,

_____ has the warmest temperature at _____ °F.

_____ has the coldest temperature at _____ °F.

The difference between these two temperatures is _____ °F.

3. On the map,

_____ has the warmest temperature at _____ °F.

_____ has the coldest temperature at _____ °F.

The difference between these two temperatures is _____ °F.

LESSON 10·6 Math Boxes

1. Draw and solve.

Mateo wants to read 8 books.

He has read 2 books.

How many more books does Mateo have to read?

_____ books

2. Sunglasses cost $3.99.

I pay $5.00.

How much change do I get back?

Fill in the circle next to the best answer.

○ **A.** $2.99   ○ **B.** $8.99

○ **C.** $8.00   ◉ **D.** $1.01

3. Write <, >, or =.

10 + 23 $\boxed{<}$ 40

18 + 5 $\boxed{=}$ 5 + 18

32 $\boxed{>}$ 51 − 20

Half of 50 $\boxed{>}$ 25

MRB 9

4. Complete the puzzle.

168		170
	179	
	189	
198		200

5. Write the number that is 10 less.

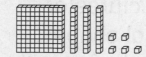

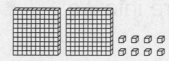

_____   _____

MRB 10–11

6. Fill in the missing numbers.

Rule

+100

 284

84   184   284   384   464

MRB 98–99

Date _____

1. $2.00 =

_____ pennies

_____ nickels

_____ dimes

_____ quarters

MRB
88–89

2. Which are you more likely to grab?

black or white? _____

○ or □? _____

MRB
47–48

3. Draw the hands to show 10:45.

MRB
80–81

4. What is the temperature? Fill in the circle next to the best answer.

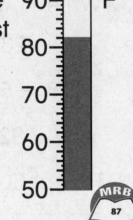

○ **A.** 82°F

○ **B.** 85°F

○ **C.** 80°F

○ **D.** 90°F

MRB
87

5. Divide the square into $\frac{1}{4}$s.

Shade $\frac{2}{4}$.

Write another name for $\frac{2}{4}$.

_____

MRB
13

6. These are pictures of 3-dimensional shapes. Put an X on shapes with curved faces.

cone   sphere   prism   pyramid

MRB
58

LESSON 10·8 Math Boxes

1. What time is it?

_____ : _____

MRB
80–81

2. Write the amount.

52 ¢

3. What day is it today?

_____

What day will it
be tomorrow?

_____

4. Complete the number-grid puzzle.

	125	
134	135	136
144	145	146

5. Count by 2s.

36, 38, 40,
42, 44, 46,
48, 50, 52,
54, 56, 58

6. What is the temperature?

_____ °F

°F
90
80
70
60
50

MRB
87

Notes

Date _____ Time _____

Notes

_____

_____

_____

_____

_____

_____

_____

_____

_____

_____

_____

_____

_____

_____

_____

_____

_____

_____

LESSON 6·7   Fact Triangles 4

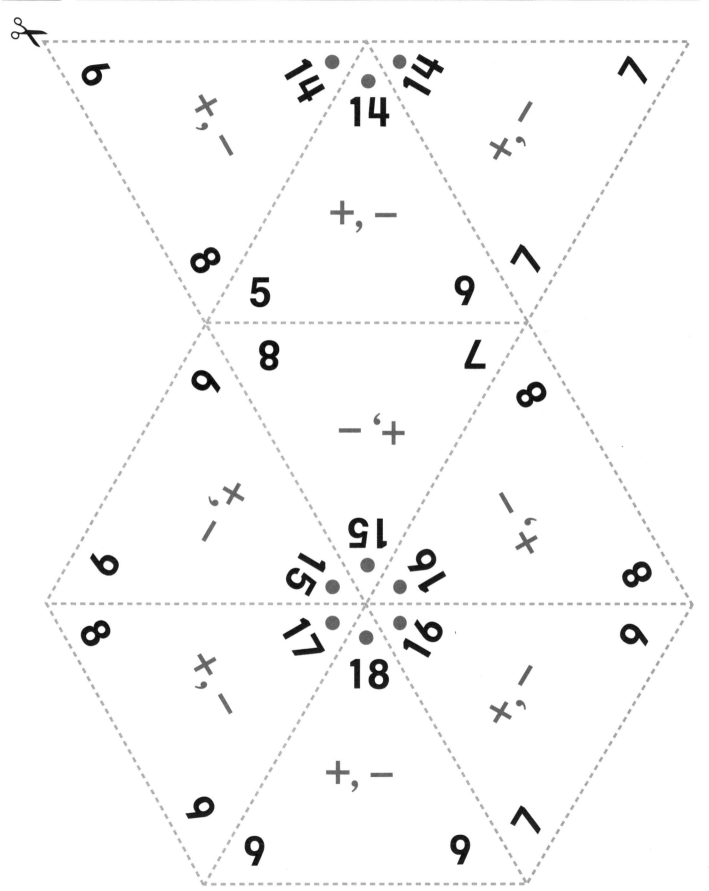

Base-10 Pieces

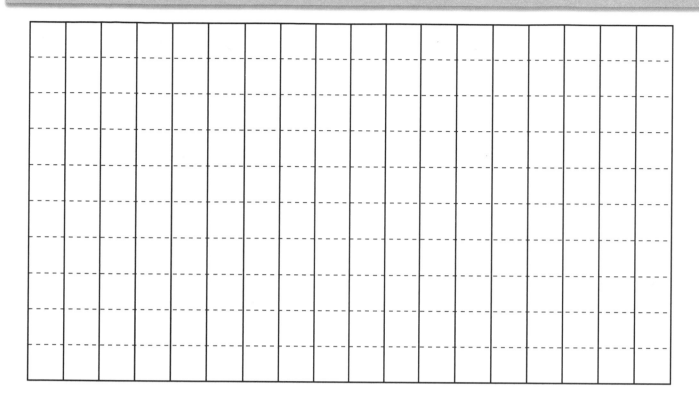

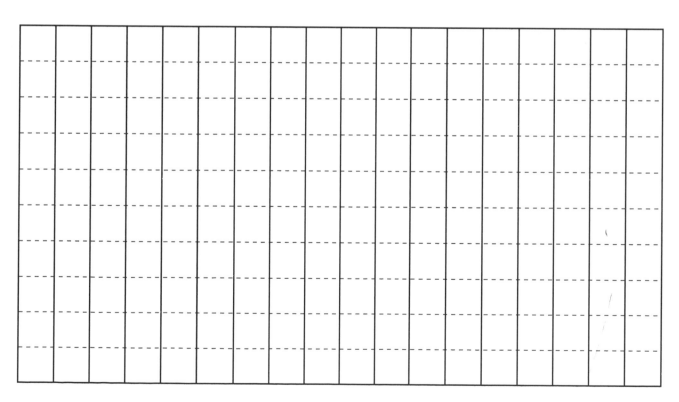

Base-10 Pieces

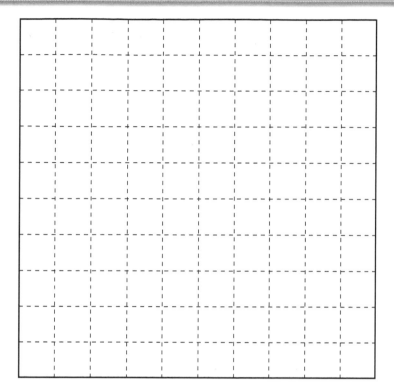

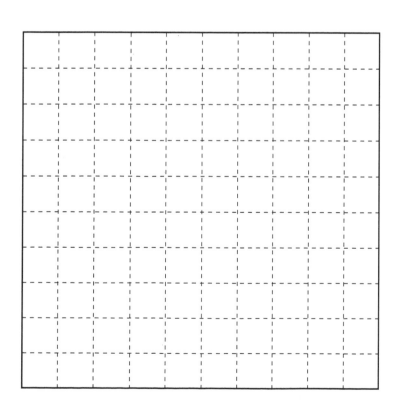

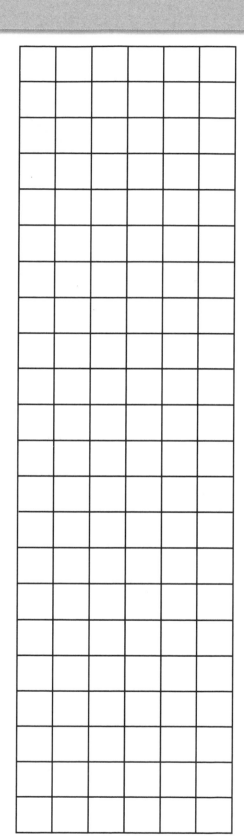